AFRICAN AMERICAN SOLDIERS

AFRICAN AMERICAN SOLDIERS

EDITED BY JOANNE RANDOLPH

PIONEERING AFRICAN AMERICANS

Enslow Publishing
101 W. 23rd Street
Suite 240
New York, NY 10011
USA
enslow.com

This edition published in 2018 by
Enslow Publishing, LLC.
101 W. 23rd Street, Suite 240
New York, NY 10011

Additional materials copyright © 2018 by Enslow Publishing, LLC

All rights reserved.

No part of this book may be reproduced in any form without permission in writing from the publisher.

Library of Congress Cataloging-in-Publication Data

Names: Randolph, Joanne, editor.
Title: African American soldiers / edited by Joanne Randolph.
Description: New York, NY : Enslow Publishing, 2018. | Series: Pioneering African Americans | Includes bibliographical references and index. | Audience: Grades 5–8.
Identifiers: LCCN 2017022379| ISBN 9780766092532 (library bound) | ISBN 9780766093973 (pbk.) | ISBN 9780766093980 (6 pack)
Subjects: LCSH: African American soldiers—History. | United States—Armed Forces—African American troops.
Classification: LCC E185.63 .A37 2018 | DDC 355.0089/96073—dc23
LC record available at https://lccn.loc.gov/2017022379

Printed in the United States of America

To Our Readers: We have done our best to make sure all website addresses in this book were active and appropriate when we went to press. However, the author and the publisher have no control over and assume no liability for the material available on those websites or on any websites they may link to. Any comments or suggestions can be sent by email to customerservice@enslow.com.

Photos Credits: Cover, p. 3 PhotoQuest/Archive Photos/Getty Images; pp. 8, 26 Buyenlarge/Archive Photos/Getty Images; p. 9 © North Wind Picture Archives; p. 10 U.S. Army; p. 12 Afro American Newspapers/Gado/Archive Photos/Getty Images; pp. 15, 30, 32 Bettmann/Getty Images; p. 17 Coover/Library of Congress/Corbis Historical/Getty Images; pp. 18, 35 Library of Congress Prints and Photographs Division; pp. 20–21 De Agostini Picture Library/Getty Images; p. 22 Archive Photos/Getty Images; p. 25 Rainer Lesniewski/Shutterstock.com; p. 27 Corbis Historical/Getty Images; p. 37 Chicago History Museum/Archive Photos/Getty Images; p. 39 Rex Hardy Jr./The LIFE Picture Collection/Getty Images; p. 40 PhotoQuest/Archive Photos/Getty Images; p. 42 Photo12/UIG/Getty Images; p. 44 Chip Somodevilla/Getty Images; interior pages Eugene Berman/Shutterstock.com (flag and dog tags).

Article Credits: Vicki Hambleton, "Blacks in the Military," *Footsteps*; Lisa Clayton Robinson, "Peter Salem, American Hero!" *Footsteps*; Peter H. Wood, "African Americans Bearing Arms," *Footsteps*; Eric Arnesen, "Fighting for Freedom," *Footsteps*; Richard L. Mattis, "Henry O. Flipper," *Cobblestone*; Eric Arnesen, "Exclusion and Segregation," *Footsteps*; Therese DeAngelis and Gina DeAngelis, "Silent Gun," *Cobblestone*; Eric Arnesen, "Barack Obama," *Cobblestone*.

All articles © by Carus Publishing Company. Reproduced with permission.

All Cricket Media material is copyrighted by Carus Publishing Company, d/b/a Cricket Media, and/or various authors and illustrators. Any commercial use or distribution of material without permission is strictly prohibited. Please visit http://www.cricketmedia.com/info/licensing2 for licensing and http://www.cricketmedia.com for subscriptions.

CONTENTS

CHAPTER ONE
A HISTORY OF SERVICE 6

CHAPTER TWO
AFRICAN AMERICANS DURING
THE REVOLUTION 14

CHAPTER THREE
AFRICAN AMERICANS BEARING ARMS   19

CHAPTER FOUR
FIGHTING FOR FREEDOM 24

CHAPTER FIVE
HENRY O. FLIPPER AND OTHER
WEST POINT GRADUATES 31

CHAPTER SIX
CONTINUED EXCLUSION
AND SEGREGATION 36

CHAPTER SEVEN
THE FIRST AFRICAN AMERICAN
COMMANDER IN CHIEF 43

GLOSSARY 46
FURTHER READING 47
INDEX 48

CHAPTER ONE

A HISTORY OF SERVICE

African Americans have participated in every war in which the United States has been involved. While their accomplishments were not always officially recognized or documented, their courage and valor did not go unnoticed. The accounts that follow detail their patriotism and heroism.

FIRST SOUTH CAROLINA VOLUNTEERS

The Civil War began in 1861, pitting the South, or the Confederacy, against the North, known as the Union. As the powerful Union navy extended its command over southern waters, it soon took control of the Sea Islands that lay off the coast of Georgia. The plantation owners fled in front of the advancing Union troops, abandoning their slaves—more than eight thousand of them. The Union considered these people "contraband of war," which meant the Union would not return them to their owners.

Under the leadership of Colonel Thomas Higginson, these men, although still slaves, became the First South Carolina Volunteers, the first African American regular army regiment organized in the war. To the surprise of many white soldiers, the regiment captured the city of Jacksonville, Florida.

Higginson kept a diary and recorded the words, as they sounded to him, of Private Thomas Long, an ex-slave. "If we hand't become sojers, all might have gone back as it was before. . . . But now tings can neber go back, because we have showed our energy and our courage and our naturally manhood."

THE FIFTY-FOURTH MASSACHUSETTS REGIMEN

After President Abraham Lincoln signed the Emancipation Proclamation in 1863, black men were allowed to enlist in the military. The Fifty-Fourth Massachusetts Volunteer Infantry, the first of these new regiments, included mostly free blacks, not runaway slaves. The Fifty-Fourth won renown for its heroism at Fort Wagner, South Carolina. On July 18, 1863, at 7:45 p.m., the regiment received the command to attack. When the troops moved within 200 yards (183 meters) of the fort, the Confederates opened fire.

When the flag bearer faltered, a sergeant named William H. Carney took up the flag and never dropped it—even when he was wounded. For his bravery, Carney was awarded the Medal of Honor, the nation's highest award for valor.

AFRICAN AMERICAN SOLDIERS

The Fifty-Fourth Massachusetts Volunteer Infantry attacks Fort Wagner. This attack was the second attempt to take the fort. Thanks in part to the Fifty-Fourth infantry, the attack succeeded.

A HISTORY OF SERVICE

This illustration shows a buffalo soldier drinking from a canteen while riding though the Arizona desert.

BUFFALO SOLDIERS

After the Civil War ended, the government turned its eyes westward. In 1866 Congress reorganized its troops, reducing the number of regiments and establishing the "New Army."

The US Colored Troops were disbanded, and two black cavalry regiments, the Ninth and Tenth, and two black infantry regiments, the Twenty-Fourth and Twenty-Fifth, were formed.

Their duties included building roads, escorting stagecoaches and wagon trains, and protecting homesteaders from outlaws. Soon after the Tenth Cavalry was assigned to Fort Leavenworth, Kansas, the ninety troopers defeated eight hundred Cheyenne Indians in a two-day battle. According to tradition, the Cheyenne began referring to the Tenth as Buffalo Soldiers. American Indians considered the buffalo a sacred animal, and the name "Buffalo Soldiers" represented the respect American Indians had for the regiment's fighting skills. The name eventually applied to all black troops in the West.

AFRICAN AMERICAN SOLDIERS

Henry Johnson was born in 1892 and died in 1929. Many years after his death, he was awarded the Purple Heart, in 1996, and the Distinguished Service Cross, in 2002.

WORLD WAR I

After German ships began attacking American vessels at sea in 1917, the United States entered "the Great War" that had begun in Europe in 1914. The US Congress passed the Selective Service Act, which required the creation of draft boards. Every male citizen between the ages of twenty-one and thirty-one was ordered to register. Most African Americans joined the army. The marine corps excluded African Americans, and, although the navy and coast guard allowed them, both units assigned African Americans to shovel the coal used to run the boats and to act as messengers.

When African Americans criticized the rule barring them from combat, the War Department organized two black divisions, the Ninety-Second and the Ninety-Third. The Ninety-Third consisted of National Guard troops from seven states, including the 369th from New York.

The 369th was the first unit of black combat troops to arrive in France. Its soldiers served at the front for a grueling 191 days—longer than any other unit. The night of May 14, 1918, Privates Henry Johnson and Needham Roberts heard a muffled clicking sound—the enemy with a wire cutter—and shouted to the corporal on duty.

Within minutes, a hand grenade exploded near Roberts, and Johnson became a one-man army. Despite being shot four times, Johnson managed to force the German unit to retreat.

Johnson and Roberts were awarded the Croix de Guerre ("War Cross"). They were the first Americans of any color to be given France's highest honor for bravery on the battlefield. The Germans recognized their valor as well, nicknaming them the Hellfighters. At home, the 369th was nicknamed Harlem's Own.

IN THE AIR: THE TUSKEGEE AIRMEN

Although African Americans were not allowed to train as pilots during World War I, Eugene Jacques Bullard was the first African American to become a fighter pilot in that war. He sidestepped this discrimination by serving in the French Air Corps.

Later, as a result of pressure from civil rights activists, the first black air corps unit, the Ninety-Ninth Pursuit Squadron (later called the Ninety-Ninth Fighter Squadron), was created in 1941. The black pilots were sent to Tuskegee Institute, a black college in Alabama, for training.

Many in the military doubted the program would succeed and nicknamed it the Tuskegee "experiment." Lieutenant Colonel Noel

AFRICAN AMERICAN SOLDIERS

A group of Tuskegee Airmen receive their commission to fight in World War II in January 1945.

F. Parrish, a white officer who believed in fair treatment for all, trained the men well. The first graduates became pilots in the Army Air Force as members of one of four fighter squadrons. The four units comprised the all-black 332nd Fighter Group.

On April 5, 1943, Secretary of War Henry Stimson assigned the Ninety-Ninth to North Africa and war duty after receiving a letter from first lady Eleanor Roosevelt. The pilots soon earned a reputation for their talent and bravery. In Italy, one of their most difficult assignments was to escort B-17 bombers. These planes could not move quickly because the bombs they carried might explode. So, to defend them against enemy attacks, the squadron would act as escorts, flying around the bombers in planes with red tails. The Tuskegee "Red Tails" flew a total of 1,578 combat missions, more than any other unit in Europe, and did not lose a single US bomber to enemy fire. For outstanding performance and extraordinary heroism, the unit received an official citation. Yet, despite their fame and talent, no commercial airlines would hire them as pilots and, for most, their flying came to a sudden end.

ON THE SEAS: THE GOLDEN THIRTEEN

At the start of World War II, the US Navy assigned African Americans to serve only as cooks and servants. On board the ship, they had to sleep together and could associate with whites only in the line of duty. In 1943 President Franklin Delano Roosevelt ordered the navy to accept African American draftees as well as recruits and to assign them to general duty. The president also agreed that a few black sailors should be selected for commissioning as officers.

Sixteen men, chosen by the secretary of the navy, passed the required training program. Since the original plan called for twelve black officers, only twelve were commissioned. A thirteenth was named a warrant officer. These thirteen men, who later called themselves the "Golden Thirteen," were the first black officers in the history of the US Navy. Not one of them, however, was ever assigned to duty outside the United States.

CHAPTER TWO

AFRICAN AMERICANS DURING THE REVOLUTION

The battles of Lexington and Concord and the Battle of Bunker Hill are two of the best-known battles of the American Revolutionary War. But not many people know that some of their heroes were black. Peter Salem, a former slave, was one of them. A minuteman soldier, he became one of the first heroes of the Revolutionary War. Salem was born around 1750 in Framingham, Massachusetts. His owner, Jeremiah Belknap, was from nearby Salem and named Peter after his hometown. Not much is known about Salem's early years or his family, except that his owner eventually sold him to Major Lawson Buckminster.

When the Revolutionary War began, only free blacks were allowed to enlist in the Continental Army. But many slaveholders began freeing their slaves just so they could join the militia and fight against the British. Buckminster gave Salem his freedom in exchange for enlisting. Salem joined the Massachusetts minutemen.

Salem was among the minutemen who fought in the Battle of Concord, on April 19, 1775, at the very beginning of the Revolutionary War. Two months later, Salem was with a group of

AFRICAN AMERICANS DURING THE REVOLUTION

soldiers led by Captain Luke Drury in the battle to control Boston's Bunker Hill—making him one of about thirty black soldiers to fight in that battle. Salem was rumored to have killed Major John Pitcairn, who was one of the leaders of the British forces on the hill and who had also led some of the troops fighting against Salem and his fellow patriots at Concord. Legend says that just as Pitcairn was ordering American forces to retreat from the hill, Salem boldly stepped forward and shot him. Historians now doubt whether this story is true, but they do agree that Salem was one of the heroes that day.

This engraving shows Peter Salem shooting Pitcairn during the Battle of Bunker Hill.

AFRICAN AMERICAN SOLDIERS

Salem, however, was not the only African American "Salem" to be a hero in that battle. Salem Poor, a black soldier from Andover, Massachusetts, fought so well that a group of fourteen white officers petitioned the Massachusetts General Court to reward him.

How were black patriots such as Peter Salem and Salem Poor really rewarded? Sadly, at the time, they were not rewarded at all. In fact, when the Continental Army, representing all thirteen American colonies, was formed just a few months later, the Southern colonies convinced those in charge of enlistment regulations to make it illegal for blacks to fight. The army reversed the rule only after the British began promising slaves their freedom if they fought on the British side. It was after the Continental Army allowed black men to join that Peter Salem reenlisted.

After the war, Salem moved to a farm in Leicester, Massachusetts, where he worked as a cane weaver. He died in 1816 in the poorhouse at Framingham. In 1882, more than sixty years after his death, Framingham built a monument in his honor.

PRINCE HALL

Prince Hall was the key organizer of African Americans in Boston during the American Revolution. He was probably born in Barbados and sailed to Boston when he was a teenager. A leatherworker, Hall sold drumheads to the British army during the war. Hall was also the founder of the first lodge of black Freemasons, a role that helped him lead other free blacks in their struggle for civil rights. He wrote several petitions, or requests, to the Massachusetts government asking it to outlaw slavery, provide free schools for black children, and pay for blacks to emigrate to Africa.

AFRICAN AMERICANS DURING THE REVOLUTION

This photo, taken in 1897, shows a lodge of black Freemasons. There are still African American lodges of Freemasons today.

CRISPUS ATTUCKS

Though not exactly a soldier, Crispus Attucks was the first person to die in the American Revolution. He probably escaped from slavery in his mid-twenties and became a sailor on a whaling ship. Attucks worked on the docks in Boston before the Revolutionary War. At

AFRICAN AMERICAN SOLDIERS

This memorial portrait of Crispus Attucks was created to remind people of his untimely death during the Boston Massacre.

the time, the city was a hotbed of tension between the local townspeople, who blamed the British government for unfair taxes, and the British soldiers patrolling the streets.

On the evening of March 5, 1770, Attucks and a group of men confronted a gathering of soldiers at the Customs House, the building where the British collected taxes. In the confusion that followed, the soldiers shot and killed Attucks and four other townspeople. The incident became known as the Boston Massacre and Attucks "the first to defy, the first to die."

CHAPTER THREE

AFRICAN AMERICANS BEARING ARMS

Blacks who took part in the American Revolution were not the first people from Africa to bear arms in North America. African newcomers to mainland North America, both enslaved and free, had been shouldering weapons for two centuries. They fought for the European colonizers, took part in Colonial Indian wars, and fought against whites to obtain their own freedom.

SETTING FOOT IN THE NEW WORLD

Africans also accompanied early Spanish explorers as soldiers and servants. When a large expedition met disaster along the Gulf Coast in the late 1520s, one of the four survivors was a Spanish-speaking African named Esteban. The four eventually made their way back to Mexico City. Later, Esteban was killed by Zuni Indians while guiding a Spanish party in New Mexico.

When the Spanish explorer Hernando De Soto invaded the southeast of what is today the United States in the 1530s, Africans

AFRICAN AMERICAN SOLDIERS

were part of his army. Records show that one was a West African named Gomez. Gomez escaped from the Spanish expedition in South Carolina and married an American Indian leader, the queen of Cofitachequi, an Indian province.

By the 1620s, Africans were also present in English and Dutch settlements on the Atlantic coast. Since the small communities lacked manpower for defense, African men were expected to train

This 1847 painting shows Hernando de Soto as he discovers the Mississippi River in 1541. Some of the people in the image appear to be Africans.

AFRICAN AMERICANS BEARING ARMS

with the militia. In the Massachusetts colony, an early muster roll from Plymouth lists a "blackamoor" (an old English term for an African). In New Amsterdam (present-day New York City), slaves serving the Dutch West India Company built fortifications and joined in fighting Indian and English enemies. Spanish-speaking blacks and mulattos (people having one black and one white parent) in Spanish New Mexico and Florida also served as soldiers.

BLACK SOLDIERS IN THE AMERICAN SOUTH

Blacks in the early colonies became skilled in the use of guns, especially since firearms were needed to hunt game and to guard livestock from wild animals. A 1703 South Carolina law offered ten shillings to any slave who killed a wolf, panther, or bear. The next year, the assembly passed an act for enlisting slaves "in Time of Alarms." By 1708 the colony's militia consisted of nearly 1,900 men, almost half of whom were enslaved blacks. For each white person in the militia, captains were also obliged by law to "train up, and bring into the field . . . one able slave armed with a gun or lance."

As slave numbers increased, southern colonies passed laws to limit black workers from possessing firearms and to prevent them from running away to join with American Indians. During North Carolina's Tuscarora War in 1711–1712, a white commander faced a cleverly protected Indian fort. To his dismay, he

AFRICAN AMERICAN SOLDIERS

This map shows the original thirteen colonies, as well as which countries control the other visible areas. Florida was under Spanish control at the time.

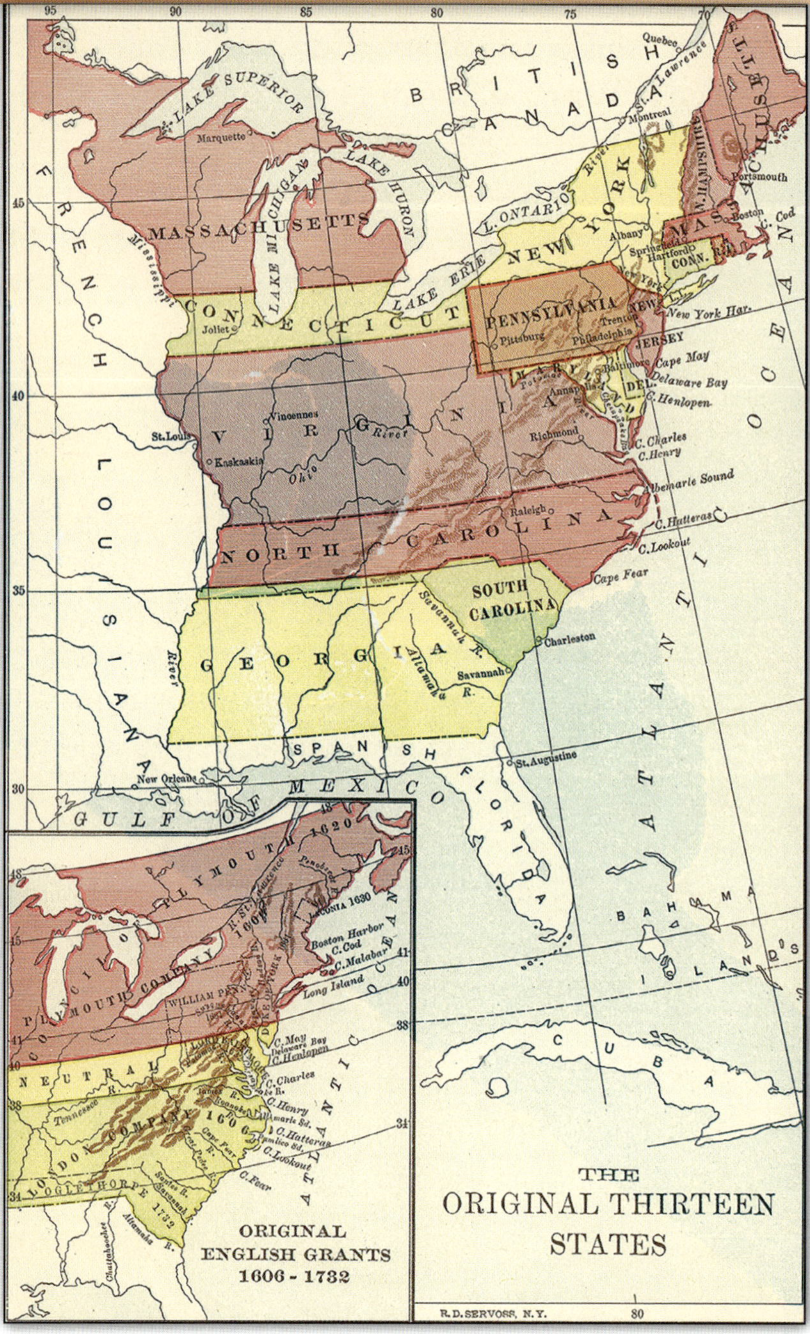

discovered that "it was a runaway negro [who had] taught them to fortify thus, named Harry."

Carolina slaves also escaped to St. Augustine in Spanish Florida, where they formed a black militia company at nearby Fort Mose and fought against the English. In South Carolina, nine months after the suppression of the 1739 Stono Revolt, several hundred slaves failed in a plan "to break open a store-house, and supply themselves, and those who would join them, with arms."

In the North, in the 1700s, where the proportion of slaves was lower and the threat of insurrection was less, blacks continued to serve in militia companies.

George Gire of Massachusetts, for example, was among the free black men who earned a pension after the French and Indian War (1754–1763). Black boys and men frequently served as drummers for military units; others worked as harbor pilots for military vessels. African American colonists also went to sea, taking part in privateering voyages or serving in the British navy.

CHAPTER FOUR

FIGHTING FOR FREEDOM

When the Civil War broke out in 1861, few white Americans would have anticipated that African Americans would serve in the US military. For many white people in the North, the Civil War was a war to "save the Union," a conflict limited to the narrow goal of reuniting the North and the South.

The "paramount object in this struggle," President Abraham Lincoln declared in 1862, "is to save the union, and is not either to save or destroy slavery." In many people's eyes, this was a "white man's war" that would end when the slave states of the new Confederate States of America rejoined the United States.

NO RESPECT IN THE NORTH

The idea of black military participation and the outright abolition of slavery won little public approval among northern whites in the first year of the war. Yet large numbers of African American men in the North demanded the right to take up arms on the side of the United

FIGHTING FOR FREEDOM

This map shows which states allowed slavery and which ones did not at the beginning of the Civil War.

States. When the war began, black men in cities across the North enthusiastically formed informal rifle companies and attempted to join the army. To their surprise, the US government rejected their offers. That policy would quickly change.

As the war dragged on, white casualties mounted, enlistment rates for new soldiers fell, and morale wavered. Some whites came to believe that enlisting black soldiers would spare the lives of white men and strengthen the Union effort. They also came to understand that only by attacking slavery in the South could the Union be victorious.

The vigorous lobbying and education campaigns waged by "Emancipation Leagues" made up of Northern blacks and their white abolitionist allies also helped make the end of slavery and the

25

AFRICAN AMERICAN SOLDIERS

enlistment of black soldiers a political reality. In mid-1862, the US Congress voted to give Lincoln the option of using black troops. In his Emancipation Proclamation, which took effect on January 1, 1863, Lincoln took that option, officially authorizing the enlistment of black soldiers.

The war to save the Union had become a war to end slavery, and black soldiers would play a vital role. From 1863 to the war's end in 1865, roughly 186,000 African Americans took up arms against the Confederacy and the system of slavery it fought to uphold. More than one third of these men, some 68,178, would die. Who were these soldiers? Fifty-three thousand were Northern free blacks opposed to slavery and determined to take part in its destruction.

The Emancipation Proclamation changed the federal legal status of millions of slaves in the rebelling Southern states to free people if they could escape Confederate control.

FIGHTING FOR FREEDOM

Company E of the Fourth United States Colored Troops is just one of the all-black units that were formed during and after the Civil War.

Approximately forty thousand came from the border states, slave states that did not secede from the Union and join the Confederacy. The largest number, ninety-three thousand, came from the slave states of the Confederacy. These escaped slaves saw the war as a way of winning freedom for themselves and their families and friends.

Fighting to free the slaves and winning the right to serve in the Union military did not mean that black soldiers had won equality for themselves. Until late in the war, they received only three-fifths of the pay of white soldiers. They served in segregated, all-black units—the United States Colored Troops—led by white officers.

AFRICAN AMERICAN SOLDIERS

At times, white Union recruiters rounded up escaped slaves in the South and forced them to join the army. In some cases, they were assigned not to combat but to work details building fortifications.

Those who did fight faced greater risks than white soldiers. To the officers of the Confederate army, black Union soldiers were fugitive slaves, not soldiers. They refused to treat captured black soldiers according to the rules governing prisoners and sentenced them to slavery or death.

THE US COLORED CAVALRY

Before the Civil War, many whites believed that blacks were too ignorant and cowardly to make good soldiers. But it did not take long for African American soldiers to prove their worth. Officers and fellow soldiers often praised the African Americans' performance on the battlefield. Northern white soldiers, once they fought next to black soldiers, came to accept them as valued partners.

In December 1863, the first two regiments of African American cavalry, designated the First and Second US Colored Cavalry (USCC), were organized in Virginia. They participated in combat operations as part of the force under Major General Benjamin Butler in eastern Virginia until the war ended. The Third USCC, created in March 1864, operated primarily out of Vicksburg, Mississippi. The Fourth USCC, formed in New Orleans in April 1864, protected various sites in southern Louisiana. The Fifth and Sixth USCC were raised at Camp Nelson, Kentucky, in October 1864.

They were involved in various actions in Kentucky and southwestern Virginia. The experiences of the Fifth USCC on a campaign in October 1864 illustrate what military service was like for many blacks. Recently recruited, the Fifth was not yet fully organized, and most of its officers had not been appointed. The troops and their horses were poorly trained. Their weapons were infantry rifles, useless to soldiers on horseback. At first, the white soldiers with whom the African Americans marched ridiculed and insulted the black cavalrymen.

Union troops attacked an enemy line at Saltville, Virginia, on October 2. The Confederates became enraged when they saw the black units fighting against them, and so they targeted those cavalrymen. But the African Americans did not falter.

Ultimately, the Union army withdrew from Saltville, leaving behind their wounded. Some of the Confederate soldiers combed the battlefield and murdered the blacks who lay hurt and helpless.

A week later, several Southern soldiers forced their way into an army hospital and murdered any black soldiers they found there. When Confederate officers heard of these atrocities, they expressed outrage and sought to identify and punish those responsible.

The Saltville battle shows the awful situation in which African Americans were placed when they tried to carry arms for their country. Many Southerners still looked upon them as slaves who needed to be "taught a lesson."

In December 1864, the Fifth USCC returned to Saltville, in company with the Sixth USCC and other cavalry. This time, they succeeded in destroying the local saltworks, depriving the Confederates of this precious commodity.

African Americans proved their worth throughout the Civil War, building a reputation as fine troopers who were courageous, able, and devoted. In 1866, when Congress scaled down the infantry,

AFRICAN AMERICAN SOLDIERS

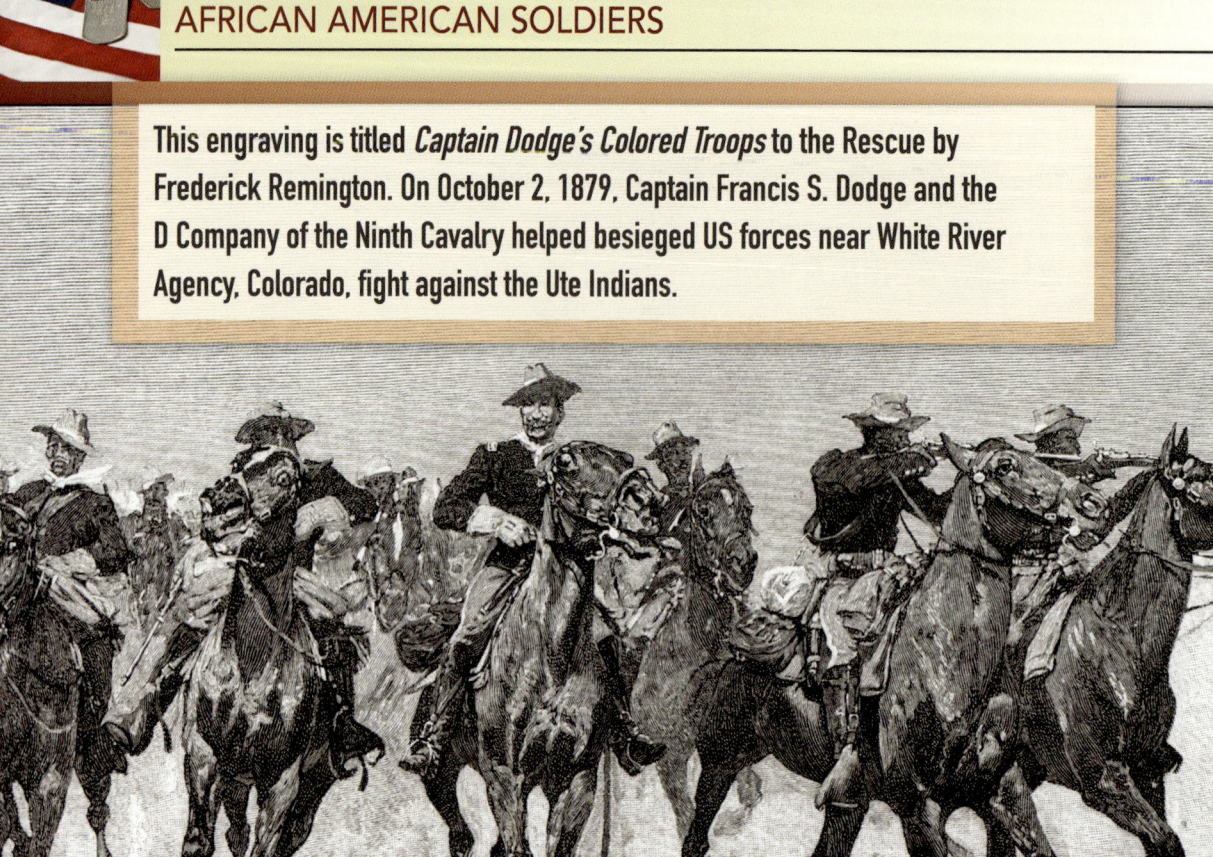

This engraving is titled *Captain Dodge's Colored Troops* to the Rescue by Frederick Remington. On October 2, 1879, Captain Francis S. Dodge and the D Company of the Ninth Cavalry helped besieged US forces near White River Agency, Colorado, fight against the Ute Indians.

it provided for ten regiments of cavalry. Two of them, the Ninth and Tenth, were all black. These units saw service in the West, primarily "subduing" American Indians who attacked settlers trying to take over Indian land.

Despite continued prejudice, African American cavalrymen made significant contributions to the settlement of the West, just as they shared in the preservation of the Union during the Civil War.

CHAPTER FIVE

HENRY O. FLIPPER AND OTHER WEST POINT GRADUATES

An unwritten rule at the US Military Academy at West Point, New York, prevented any white cadet from talking to Henry Ossian Flipper. Any cadet who spoke to him risked being shunned in the same way that Flipper was. Flipper endured four years of isolation and loneliness at the academy and, on June 15, 1877, became the first African American to graduate from West Point.

A LIFETIME OF ACCOMPLISHMENTS

Flipper was born to slave parents in Thomasville, Georgia, in 1856. A slave mechanic taught him to read and write when he was eight years old. In 1873 he applied for admission to West Point. He became the fifth African American cadet to attend the academy.

As a new cadet, Flipper thought he had found some friends in the other first-year students. But all he got from them were racial

AFRICAN AMERICAN SOLDIERS

Henry O. Flipper is shown as a young man in his cadet's uniform.

slurs and other forms of disrespect. Some cadets tried to avoid standing next to Flipper in field drill or sitting next to him in chapel. During a cavalry drill, a rider once kicked Flipper's horse with his spurs, making the horse bolt.

Flipper rose above such meanness. He made sure his own behavior was always proper, and he did not return the insults. In time, Flipper's hard work and patience slowly earned him the respect of some cadets. At graduation, the entire cadet corps applauded him, and many cadets congratulated him and shook his hand.

After graduation, Second Lieutenant Henry Flipper joined the Tenth Cavalry on the western frontier. He arrived at his first post, Fort

HENRY O. FLIPPER AND OTHER WEST POINT GRADUATES

Sill, Indian Territory (now Oklahoma), in January 1878. Flipper enjoyed the respect of the white officers and the challenge of his duties for nearly three years. He rode in a campaign against the Apaches and tracked down ammunition thieves. The young officer supervised projects to drain stagnant ponds, which were suspected of causing malaria. His troop also put up telegraph wires and built a road.

In 1880 he took charge of the commissary (a store for military personnel) at Fort Davis, Texas. He bought and sold supplies, kept track of money, and filed reports. Unwisely, he sometimes permitted soldiers to buy items on credit. He also kept cash and checks in a locked trunk in his quarters rather than in an office safe.

In July 1881, Flipper discovered a shortage in the commissary money he was responsible for keeping. Believing he could make up the shortage himself, he filed three false reports. When his commanding officer, Colonel Shafter, asked about the funds, a frightened Flipper lied to him.

Shafter accused Flipper of embezzlement and formally charged him, even though friends had loaned Flipper money to cover the loss. Some people believe that several white officers, angered by his friendship with a white woman (Flipper sometimes went riding with the sister of his captain's wife), had plotted to get rid of him. The military court found Flipper innocent of taking money but guilty of conduct unbecoming an officer. He was dishonorably discharged on June 30, 1882.

With his career as a soldier over, Flipper worked as a surveyor, mining consultant, and land claims investigator. Later, he became friendly with New Mexico businessman Albert Fall. After Fall became a US senator, he called Flipper to Washington in 1919 to serve as a consultant. Fall then became secretary of the interior and made Flipper an assistant.

AFRICAN AMERICAN SOLDIERS

Flipper tried several times to have Congress overturn his conviction and restore him to military duty, but he failed. In 1930 he moved to Atlanta, Georgia, where he lived with his brother Joseph until his death in 1940.

About thirty years later, schoolteacher Ray MacColl and Flipper's niece Irsle King asked a special committee of the US Army to review Flipper's case. In 1976 the committee agreed that the punishment had been too severe. The committee granted Flipper an honorable discharge effective June 30, 1882.

OTHER WEST POINT PIONEERS

Only two other African Americans graduated from West Point before 1936. John Hanks Alexander was born on January 6, 1864 in Helena, Arkansas. He graduated from West Point in 1887 and served in the Ninth Cavalry for most of the next seven years. In January 1894, he began teaching at Wilberforce University in Ohio. He died unexpectedly about two months later at age thirty.

Charles Young was born March 12, 1864, in MaysLick, Kentucky. His graduation in 1889 launched a distinguished military career. He served with the Ninth and Tenth Cavalry units and taught at Wilberforce University. Young later commanded troops in the Philippines and was a military attaché in Haiti and Liberia (in West Africa). He led a cavalry squadron in Mexico with General John J. Pershing in 1916 and 1917.

Shortly after the United States entered World War I, a physical examination showed that Young had nephritis, a kidney disease. The army forced him to retire, then promoted him to colonel, calling

HENRY O. FLIPPER AND OTHER WEST POINT GRADUATES

Major Charles Young was born in 1864 to enslaved parents who escaped from slavery shortly after Young's birth. His father joined the Union forces during the Civil War.

him back to active duty just before the war ended. Some people believe that the army deliberately denied Young a command in Europe because he was African American.

Young died in 1922, while serving as an advisor to the government of Liberia.

35

CHAPTER SIX

CONTINUED EXCLUSION AND SEGREGATION

As explained in the previous chapters, everyday life for an African American soldier before 1948 included harsh discrimination—and most of it was legal. Until President Harry S. Truman called for an end to discrimination and segregation that year in the US armed forces, African American soldiers who fought for their country did so under particularly difficult circumstances. Often trained apart from whites and assigned to all-black units, they frequently served under white, not black, commanders. In addition, black soldiers were, on many occasions, assigned difficult and unpleasant jobs, barred from certain better and higher-ranking positions, and treated more harshly than whites. Yet, despite these hardships, many African Americans served their country by joining the armed forces, hoping to win respect, dignity, and their rights through patriotic military service.

According to the US government, World War I aimed to make the world safe for democracy. African Americans at the time did not hesitate to point out that they were denied democracy at home. In 1918 civil rights leader W. E. B. Du Bois called on African

CONTINUED EXCLUSION AND SEGREGATION

Americans to "close ranks," support the war effort, and "have courage and determination" that "out of this war will rise an American Negro with the right to vote and the right to work and the right to live without insult."

Like Du Bois, many African Americans hoped that their country would recognize their dedicated service and that their postwar prospects would be brighter. Events proved them wrong. In the South, white mobs, who feared the very changes that black soldiers anticipated, killed black soldiers. The summer of 1919 saw brutal race riots that pitted angry whites against blacks in dozens of cities and towns.

A crowd of African Americans gather in the streets during the 1919 Chicago Race Riots.

AFRICAN AMERICAN SOLDIERS

THE DOUBLE V CAMPAIGN DURING WORLD WAR II

African Americans approached the question of black participation in World War II with fewer illusions. Once again they confronted widespread discrimination, segregation, and exclusion in war industries and the military. This time, however, they refused to "close ranks" behind the war and put aside their struggles for civil rights. Instead, they conducted a "Double V Campaign," launched by a black weekly newspaper, the *Pittsburgh Courier*. "Double V" stood for "victory over our enemies from without" (the Axis powers, mainly Germany and Japan) and "victory over our enemies from within" (racism at home).

Activist and union leader A. Philip Randolph took the most dramatic step. He threatened a march on Washington by one hundred thousand African Americans in 1941 unless the government promised to end discrimination in wartime jobs and the military. President Franklin D. Roosevelt pressured Randolph to call off the march. When Randolph refused, Roosevelt compromised, issuing an executive order forbidding racial discrimination by military contractors. Randolph had scored a huge victory. Even though much discrimination remained and the government had few powers to stop it, this was the first time since Reconstruction (1866–1877) that the government had committed itself—in theory, at least—to supporting some black rights.

But Roosevelt's executive order did not address the problem of discrimination in the military. To blacks' calls for integration in the armed forces, the War Department explained that its policy was "not to intermingle colored and white enlisted personnel in the same regimental organizations." To "make changes would produce situations

CONTINUED EXCLUSION AND SEGREGATION

destructive to morale and detrimental to the preparations for national defense." White military leaders, like many other whites, believed that African Americans were inferior to whites. Segregation, they believed, must remain firmly in place. During World War II, it did.

The marines initially barred all African Americans from their ranks. Until mid-1942, the navy enlisted African Americans only as stewards, cooks, and messmen (dining room waiters). Black soldiers were often assigned service work building roads and loading and unloading ships. When posted to military bases in the South for training, black soldiers lived in segregated housing and suffered attacks from local whites as well as white military police. The army even insisted that blood donated by blacks would be segregated from blood donated by whites on the grounds that white military men "would refuse blood plasma if they knew it came from Negro veins."

Despite the opposition of many white military commanders, the integration of the armed forces became a powerful political issue in the 1940s. As African Americans repeatedly pointed out, it

A. Phillip Randolph organized the first African American labor union, the Brotherhood of Sleeping Car Porters.

AFRICAN AMERICAN SOLDIERS

was hypocritical for the United States to be fighting Nazi racism in Europe when it practiced its own racism at home. It was also hypocritical to ask African Americans to fight for democracy abroad when it was denied them at home.

After World War II, President Harry Truman appointed a high-level committee that included blacks and whites to examine the situation of African Americans. As a new Cold War between the United States and the Soviet Union intensified, the issue of American racism became even more important. Both the Americans and the Soviets sought to win the allegiance of newly independent nations in Africa and Asia, a task made more difficult for the United States because of its poor record on civil rights. Although this did not lead the US government to embrace equal rights for African Americans, it did make political leaders somewhat more sensitive to racial issues.

The Ninety-Second Division was assigned the job of clearing mines off a beach in Viareggio, Italy, in 1944.

CONTINUED EXCLUSION AND SEGREGATION

Black leaders kept the pressure on. In 1948 the issue came to a head. As Congress debated whether to create a peacetime military draft, Randolph declared segregation in the armed forces unacceptable. "I personally will advise Negroes to refuse to fight as slaves for a democracy they cannot possess and cannot enjoy," he boldly declared in testimony before the Senate. Although many white Americans—as well as some black Americans—opposed Randolph's tactics, many African American leaders endorsed Randolph's demand that military segregation had to end.

SILENT GUN

John Gray, of Mobile, Alabama, was not welcome in the US Marine Corps (USMC) until after June 1942, when it admitted black recruits for the first time in its 167-year history.

Nineteen-year-old Gray was drafted and assigned to the Fifty-First Defense Battalion. As a member of the only black marine unit that underwent extensive combat training, he became qualified as a sharpshooter. Although his unit shipped out to American Samoa in the Pacific Ocean in February 1944, Gray and his fellow soldiers never saw any combat in the nineteen months they spent overseas.

Brigadier General Benjamin O. Davis was the highest-ranking African American in the armed forces, and he brought attention to the nearly impossible challenge of maintaining "high morale in a community that offers [black servicemen] nothing but humiliation and mistreatment." In 1941, on his recommendation, the army activated the 366th Infantry Regiment, the first all-black Regular Army unit, with all black officers. Soon after, Alabama's Tuskegee Institute of the US Army Air Corps began training the first black pilots.

AFRICAN AMERICAN SOLDIERS

Brigadier General Benjamin O'Davis takes a tour of France in the summer of 1944

It was not until 1948, three years after the war ended, that President Truman issued an executive order declaring "there shall be equality of treatment and opportunity for all persons in the armed services without regard to race, color, religion, or national origin."

The US Navy had begun integrating its forces by that time, and the US Air Force completed an integration plan four years later. Segregation in the army, however, began to break down only in the 1950s, after white units serving in the Korean War (1950–1953) suffered such huge casualties that they had to draw on black recruits to replenish their units.

During World War II, more than 2.5 million African American men and women registered for the armed forces, but fewer than one million actually were called upon to serve. The discrimination and segregation blacks faced at home carried over into military life. Still, African Americans proved their bravery and loyalty during the war—and after.

42

CHAPTER SEVEN

THE FIRST AFRICAN AMERICAN COMMANDER IN CHIEF

To almost every observer of the 2008 presidential campaign, Barack Obama's election was a historic event. For the first time, American voters chose an African American to be their nation's leader. A relative newcomer on the national political scene, Obama offered a message of hope and a promise to unify the country. That message carried him to the White House, and he was inaugurated before a record-setting crowd of two million people.

Though not a soldier per se, the President of the United States is also the Commander in Chief of the US Armed Forces. After hundreds of years struggling for the right to fight for the country, an African American was now in charge of the nation's substantial military resources. Upon entering office, Obama inherited two unpopular wars in Afghanistan and Iraq. He wound down the conflicts by withdrawing US troops and relying instead upon military contractors and using drones. Both countries remained unstable and subject to armed uprising.

In May 2011, Obama announced that Osama bin Laden, the leader of Al Qaeda, had been located and killed by US Special Forces. Al Qaeda was the Islamic group that claimed responsibility

AFRICAN AMERICAN SOLDIERS

Barack Obama became the first African American commander in chief on January 20, 2009, when he was sworn in as the forty-fourth US president.

THE FIRST AFRICAN AMERICAN COMMANDER IN CHIEF

for the September 2001 terrorist attacks on the World Trade Center in New York City and the Pentagon in Washington, DC. Although the United States is still struggling with conflicts in the Middle East, Barack Obama did his best to disentangle as many US forces and resources as he could during his time in office, but also was decisive in making strategic attacks against perpetrators of terror in those regions.

African Americans have proved time and again that they are brave and courageous patriots who deserve to stand side-by-side with other US soldiers in the fight for freedom around the globe.

GLOSSARY

abolitionist Someone who wants to get rid of slavery.
atrocities Appalling conditions or behaviors.
cavalry Soldiers who fight on horseback.
Cold War Hostility and sharp conflict between the United States and the Soviet Union, without actual warfare.
discrimination The unfair treatment of people, especially based on race, religion, or gender.
hypocritical Professing beliefs or values one does not have or practice himself or herself.
insurrection A revolt against the civil, or recognized, authorities.
minuteman An American citizen who, at the time of the Revolution, volunteered to be ready at a minute's notice.
muster roll A list of persons in a military unit.
regiment A unit of the army or armed forces that can be further divided into smaller groups, such as companies or battalions.
riots Violent acts by a group of people.
saltworks Places where salt is produced.
segregation The separation, oftentimes by race, gender, or class, of one group from another.
shilling The monetary unit used in Colonial America.

FURTHER READING

BOOKS
Lang, Matt. *Minority Soldiers Fighting in World War II*. New York, NY: Cavendish Square Publishing, 2017.
Miller, Derek L. *Minority Soldiers Fighting in World War I*. New York, NY: Cavendish Square Publishing, 2017.
Moore, Shannon Baker. *Harlem Hellfighters*. Minneapolis, MN: Essential Library, 2016.
Newsome, Joel. *Minority Soldiers Fighting in the Civil War*. New York, NY: Cavendish Square Publishing, 2017.
Reeder, Eric. *Minority Soldiers Fighting in the American Revolution*. New York, NY: Cavendish Square Publishing, 2017.

WEBSITES
CAF Red Tail Squadron
www.redtail.org/
Learn more about the history and legacy of the Tuskagee Airemen, as well current news and events.

Library of Congress
www.loc.gov/teachers/classroommaterials/presentationsandactivities/presentations/timeline/civilwar/aasoldrs/
An overview, with supporting primary source documents, of the African American experience and participation in the Civil War.

U.S. Army, Africans in the U.S. Army
www.army.mil/africanamericans/timeline.html
A timeline, profiles, and articles about African American military history.

INDEX

A
Alexander, John Hanks, 34
Attucks, Crispus, 17–18

B
Buffalo Soldiers, 9
Bullard, Eugene Jacques, 11

C
Carney, William H., 7
Civil War, 6–7, 9, 24–30

D
Davis, Benjamin O., 41
Double V Campaign, 38
Du Bois, W. E. B., 36–37

E
Emancipation Proclamation, 7, 26

F
Fifty-Fourth Massachusetts Volunteer Infantry, 7
First South Carolina Volunteers, 6–7
Flipper, Henry Ossian, 31–34

G
Gire, George, 23
Golden Thirteen, 13
Gray, John, 41

H
Hall, Prince, 16
Higginson, Thomas, 7

J
Johnson, Henry, 11

K
Korean War, 42

L
Lincoln, Abraham, 7, 24, 26

O
Obama, Barack, 43–45

P
Poor, Salem, 16

R
Randolph, A. Philip, 38, 41
Revolutionary War, 14–18, 19
Roberts, Needham, 11
Roosevelt, Franklin Delano, 13, 38

S
Salem, Peter, 14–16
segregation, 27, 36–37, 38–41, 42
slaves/slavery, 6–7, 14, 16, 17, 19–23, 24–30, 31, 41

T
Truman, Harry S., 36, 40, 42
Tuskegee Airmen, 11–13, 41

U
US Colored Cavalry, 28–30

W
World War I, 10–11, 36
World War II, 13, 38–42

Y
Young, Charles, 34–35